MEASURING UP
time

Peter Patilla

Thameside Press

Distributed in the United States by
Smart Apple Media
1980 Lookout Drive
North Mankato, MN 56003

Text copyright © Peter Patilla 1999

Illustrator: Dave Cockcroft
Editor: Claire Edwards
Designer: Simeen Karim
Picture researcher: Juliet Duff
Consultant: Martin Hollins

Thanks to Thomas and Joseph Shipman for being the readers,
and to Alison Patilla for research.

Library of Congress Cataloging-in-Publication Data

Patilla, Peter.
    Time / written by Peter Patilla.
        p. cm. -- (Measuring up)
    Includes index.
    ISBN 1-930643-17-9
        1. Time measurements--Juvenile literature. 2. Clocks and watches--Juvenile literature.
    [1. Time measurements. 2. Clocks and watches.] I. Title.

    QB209.5 .P38 2001
    529--dc21

                                                                2001027177

Printed in USA

9 8 7 6 5 4 3 2 1

Picture acknowledgments: Ancient Art & Architecture: 4, 8, 11, 15, 22, 24; **Bridgeman Art Library:** 10 Louvre, Paris; **Mary Evans Picture Library:** 9; **Werner Forman Archive:** 13, 19 bottom; **Harpur Garden Library:** 28 Marcus Harpur; **Michael Holford:** 6, 17 bottom, 19 top, 20; **Science and Society Photo Library:** 25; **Science Photo Library:** 5 Robin Scagell, 7 Smithsonian Institution, 16 top Jean Loup Charmet, 23 David Parker, 27 top Alexander Tsiaras, 27 bottom; **Tony Stone Images:** 12, 29.

CONTENTS

Thousands of years ago, the earliest civilizations noticed that the seasons had a pattern. By matching this to the movement of the sun, moon, and stars, people began to measure time, and divide it up into years and days.

The first question people must have asked about time was "when?" rather than "how long?" They needed to know when to plow and sow, and when to move their cattle. The important times were night and day, and the changes brought about by the seasons.

Even the migration of animals helped early people decide the time of year.

Archaeologists have found patterns on pebbles and bones that are thousands of years old. They think these may have been early attempts to record the passing of time.

The moon

The first way in which ancient people kept track of the passing of time may have been through looking at the changes in the moon's appearance. They could easily see how the moon moved across the night sky, and how the shape of the moon seemed to change from full moon to new moon and back to full moon again night by night. They didn't yet know that this was because the moon circles the Earth once each month.

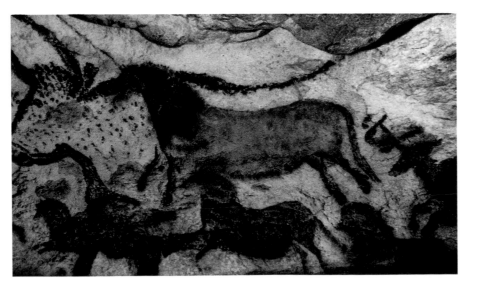

These animals were painted about 17,000 years ago in a cave in Lascaux, France. People knew by the change in the seasons when to follow the herds. They may have drawn such pictures as a kind of magic to help in their hunting.

The stars

People saw that the stars moving across the night sky followed a pattern. They also saw that certain stars appeared on the horizon just before sunrise or just after sunset at certain times of year. In this way, they learned to use the stars to measure time. The Aborigines of Australia used the appearance of the bright stars Vega and Arcturus to tell them the best times to go hunting. Other people measured time by a group of stars called the Pleiades.

The movement of the sun was often linked with early religions. Ancient farmers and cattle herders built huge stone circles, like Stonehenge, both for sun worship and to mark the seasons.

The sun

Another way to measure time was by watching how the sun seemed to move across the sky causing periods of light and dark. Early astronomers noticed that the sun rose in a different place each day over a period of time, which we now call a year. At midday on midsummer's day they noted that the sun reached its highest point in the sky. At midday in midwinter the sun reached its lowest point.

FIRST CALENDARS

The passing of a year is marked by a calendar. Thousands of years ago, early civilizations worked out their own calendars based on the movements of the sun, moon, and stars.

Early astronomers calculated that there were 360 days in a year and 24 hours in a day. But they also realized that years and months did not divide into an exact number of days. In fact, a year is 365 ¼ days, and a lunar month is just over 29 ½ days. Different civilizations balanced out the year in different ways.

This clay tablet from Babylon shows a priest's calculations for when new moons occur. This information was important as a new moon marked the start of a new month.

The Sumerians

The Sumerians, who lived more than 5,000 years ago, were the first people to divide a year into days, and days into hours. They based their calendar, which was kept by the local priest for each town, on the moon. Each year was divided into 12 lunar months of 30 days each. This made the year 360 days long. The Sumerian priests knew that their 30-day months created an error in their calculations, because the seasons occurred later and later each year.

Time Fact
A millennium is a period of 1000 years. The modern calendar begins with the year 1, and is based on the Christian date for the birth of Christ.

The Babylonians

The Babylonians based their calendar on the Sumerian calendar, and by 2000 B.C. they had improved upon it. Instead of only 30-day months, each month had either 30 or 29 days. To keep in check with the sun, they would throw in an extra 30-day month from time to time to make up for the lost days. In the same way, they added an extra month every three years. This rather hit-or-miss way continued for about 15 centuries.

The Egyptians

The Ancient Egyptian calendar was developed in about 4000 B.C. The years, which were always the same length, had 365 days in them. This calendar was based on the seasons and the flooding of the Nile River, rather than the moon. The movement of the sun and stars helped them work out a 24-hour day. Their calendar had 12 months of 30 days each, and the extra five days were spent celebrating and feasting. This calendar remained in use until about 238 B.C. when Egypt's rulers decided that an extra day should be added each fourth year—just like our modern leap year.

Calendars From Asia

Writings from India in about 1000 B.C. show that the Hindu calendar was made up of 360 days. There were 12 months, with a leap month every five years. The Ancient Chinese were also excellent astronomers. By about 1300 B.C., they had drawn up an amazingly accurate calendar made up of 365 ¼ days.

The Maya

The Mayan civilization flourished in Central America from about 2000 B.C. until about A.D. 1500. Their lives were ruled by many different calendars. The Mayan year had 18 months, each of 20 days. This left an extra five days, which the Maya thought were full of evil omen.

The Maya built observatories to help them observe the sun, moon and planets and draw up incredibly accurate calendars.

MODERN CALENDARS

The Egyptian calendar was adapted by the Greeks, and then by the Romans. Our modern calendar developed from this.

By the first century B.C. the Roman calendar, based on the moon, was in a mess. The Roman Emperor Julius Caesar invited Sosigenes, a Greek-Egyptian astronomer, to help him sort out the chaos.

The Julian calendar

Sosigenes said that a year should be 365 days long, and based on the sun and seasons. To bring the year into step with the seasons the Romans added 23 days to the end of February and 67 days between the months of November and December. The year became known as "the year of confusion". To keep the calendar accurate, Caesar ordered that every fourth February should have an extra day. This was the first modern leap year. The new calendar was called the Julian calendar.

On this Roman calendar the months and days of the week are marked by pegs.

Time fact
A decade is a measurement of time meaning ten years. It comes from the Latin word for ten.

Out of step again

After Julius Caesar's death, the Romans sometimes put the leap year every three years instead of four, although in about 8 B.C. the Roman Emperor Augustus tried to sort this out. By about A.D. 710 an English monk called Bede realized that the seasons were changing slightly earlier than they had 400 years ago. People began to add extra days to the year, but different countries, and even different towns, used different systems.

Gregorian calendar

In 1582 Pope Gregory XIII realized that over the years Easter was coming later and later. This was because a year is not exactly 365¼ days long – Sosigenes was 11 minutes out in his calculations. Over hundreds of years this made a difference of several days. The Pope ordered that 10 days be taken from October that year to put things right, but only Roman Catholic countries did this. Britain did not change to the Gregorian calendar until 1752. In this year September 2 was followed by September 14, and New Year's Day was moved from March 25 to January 1. The Gregorian calendar is now used worldwide.

Metric time

The French Revolution brought us metric measures. It nearly brought us metric time too. The government decided that from 1792 onward the years would be given Roman numerals, starting with Anno I. The calendar had 12 months, each of 30 days, with five extra days at the end of the year. They made 10-day weeks, 10-hour days, 100-minute hours and 100-second minutes. This was abandoned by 1806.

This clock from the French Revolution shows both the old time and the new 10-hours a day time.

Did you know?
In 1752 there were riots in Britain when the calendars were changed and 11 days were lost. People thought that their lives would be shorter.

MONTHS

The length of a month is about the time it takes for the moon to go around the Earth once. But have you ever wondered why the months are called what they are? As with some of our measurements, we owe the names of the months to the Ancient Romans.

When Emperor Augustus tried to sort out the calendar in about 8 B.C., he changed the order and the names of the months, as well as the number of days in some months. Augustus is responsible for the names and the order of our modern months.

Augustus Caesar was Emperor of Rome. He lived from 63 B.C. to A.D. 14. He was the great-nephew of Julius Caesar.

A 10-month year

In early Roman times there were 10 months in each year. Six were named after gods and four were given numerical names. Before 700 B.C., the year began in March. Six months had 30 days each and four had 31 days. This made the year 304 days long. The months, in order, were: Martius, Aprilis, Maius, Junius, Quintilis, Sextilis, Septembris, Octobris, Novembris, and Decembris.

Time Fact
The average time it takes the moon to go around the Earth is 29 days, 12 hours, 44 minutes, 3 seconds. This period is called a lunar month.

New months

In about 700 B.C., Numa, the king of Rome, added two extra months, called Januarius and Februarius, to the calendar. The names of the remaining 10 months stayed the same. He also changed the number of days in a year by adding an extra 50 days. He took off one day from each of the 30-day months and shared these 56 days between his new months of January and February.

Names of the months

January is named after the Roman god of doorways and beginnings. Janus had two faces, looking in opposite directions. January looks back to the old year and forward to the new year.

February is named after Februalia, a Roman festival, where people washed away their sins.

March takes its name from the god Mars. At first Mars was a god of agriculture. Later he became the god of war.

April may have been named after the Greek goddess Aphrodite, the goddess of love. It could also have come from the Latin word aperire, which means "to open", like flowers in spring.

May is named after the goddess Maia, whose name meant "mother" or "nurse". She was linked to growing crops.

June is named after the goddess Juno, wife of Jupiter and mother of Mars.

This picture from a book of hours shows the month of June.

July was once the fifth month of the year, called Quintilis (from the Latin word quintus, which means fifth). It was changed to July in honor of Julius Caesar.

August was once the sixth month of the year, called Sextilis (from the Latin word sextus, meaning sixth). It was changed to Augustus in honor of the Roman emperor Augustus Caesar.

September, October, November and December
Their names came from Roman words meaning seventh, eighth, ninth, and tenth. They held these positions in the year until Julius Caesar moved the start of the year from March to January in 45 B.C.

Mars, god of war, in his chariot.

WEEKS

Weeks have not always been made up of seven days. Hundreds of years ago, a week was probably the period of time between market days. This could have been as few as four days or as many as 10 days.

Unlike years and months, a week is not based upon any astronomical observations. The Sumerians and Babylonians divided their year into weeks of seven days, and one of these days was for resting and recreation.
The Babylonians named their days after the sun, the moon, and the five planets they had discovered at that time.

Hebrew weeks
The Ancient Hebrew people may have borrowed the seven-day week from the Babylonians, but they also based it on a text in the Bible. In the story of the Creation, God worked for six days creating Earth, and rested on the seventh.

From ancient times, market days were traditionally at the end of the working week. People came to sell their goods, but there were also musicians and other entertainers.

Roman weeks
The Ancient Romans did not divide their months into weeks. They used three days to mark different parts of each month. The first of each month was called the Calends. The fifth or seventh day was called the Nones. The third marker in the middle of the month was called the Ides. For a time, the Romans used an eight-day week, but this was for government rather than for everyday use.

Roman change
In A.D. 321 the Christian ruler Emperor Constantine introduced a new seven-day week in the Roman Empire. He ordered that the first day of the week should be Sunday, and that it should be a day of rest. Only farmers were allowed to work that day. The seven days were called Sun's day, Moon's day, Mars' day, Mercury's day, Jupiter's day, Venus' day and Saturn's day.

Weekday names

The English names we use for the days of the week come from Anglo-Saxon words for the Norse gods of the Teutonic people. Teutonic people came from Germany, Britain, Netherlands, and Scandinavia.

The Norse goddess Frigg.

Sunday *comes from the Romans and was named in honor of the sun.*

Monday *also comes from the Romans and was named in honor of the moon.*

Tuesday *was named after Tiw, a Norse god of war.*

Wednesday *was named after Woden, the most important of all the Norse gods.*

Thursday *was named after Thor, the Norse god of thunder and war.*

Friday *was named after Frigg, the Norse goddess of marriage. Frigg was the wife of Woden.*

Saturday *comes from the Romans and was named in honor of Saturn, a Roman god of agriculture.*

This Viking statue from about A.D. 1000 shows Thor, God of Thunder.

Time Fact
In the past a priest called out to announce the start of each new month. Our word calendar comes from this calling, or "calend."

DAYS AND HOURS

Counting in days is the easiest way to measure time, because it is so clearly marked by daylight and dark.

Ancient civilizations started each new day at different times. The Egyptians began their new day at midnight and the Hindus at dawn. The Babylonians, Jews, Greeks, and Muslims began theirs at dusk.

Beginning with midnight

The problem with counting the hours of a new day by starting at dawn or dusk is that they come at different times during the year. The Romans decided to start each new day at midnight, which never varies, so that the hours of day and night were equal. In this way, each day started at the same point, no matter what time of year it was.

Time Fact
In the U.S. armed forces, a new day begins at midnight (00.00) and ends at 23.59 or one minute to midnight.

Morning and afternoon

The part of the day between midnight and noon is called morning. It is given the letters A.M., which stand for *ante meridian*. *Ante* is the Latin word for "before," and *ante meridian* means before the sun reaches noon. The second part of the day between noon and midnight is called afternoon. It can be given the letters P.M., which stand for *post meridian*. *Post* means "after" and *post meridian* means after the sun has reached noon.

This sixteenth-century clock was built in Switzerland. Public clocks became more important as people began to plan their lives by the hour.

The hour

Although ancient civilizations, such as the Egyptians, had divided the day into 24 parts, people did not regulate their lives by this. They rose at dawn, ate when they were hungry, and went to bed when they were tired. The Anglo-Saxons broke the day up into chunks of time called tides, which meant season or hour. They had morningtide, noontide, and eventide. Once people began to hold markets or arrange meetings at a particular time, they needed to measure time more accurately. At first they used the position of the sun, or sundials. In the fourteenth century, with mechanical clocks coming into more common usage, people were able to plan their lives more and more by the hour.

Did you know?
In early times, the German and Scandinavian people used to count nights rather than days. It is from this that we get the word fortnight, meaning 14 nights.

Parts of an hour

Each hour can be broken down into smaller parts called minutes and seconds. There are 60 minutes in one hour and 60 seconds in one minute. The idea of breaking up an hour into 60 parts rather than 10 or 100 dates back to the Babylonians. They divided many of their measurements into 60 parts. We have just carried on what they started all those thousands of years ago.

SHADOW TIME

Thousands of years ago, astronomers divided the year according to the sun, moon and stars. But for most people, telling the time accurately was unimportant.

Without such things as public transportation or television, people did not need to know the exact time. Timekeeping was usually limited to estimates such, as early morning or late evening. Despite this, from the earliest civilizations, people did want to measure the passing of time during the day, for such things as praying and organizing religious duties.

Shadow sticks

The earliest way in which people learned to tell the time was by watching shadows. The simplest marker would have been an upright stick. The length of shadow cast by the stick showed the passing of time, from a long shadow at dawn, a short shadow at midday, and a long shadow again at dusk. Later on, people built carved obelisks (pillars) that cast huge shadows on the ground.

Shadow clocks

The Egyptians were probably the first people to make shadow clocks. They didn't look like modern clocks, but they had hours marked on them so that they told the time more accurately than before. An Egyptian shadow clock was made of two bars set at right angles like a T. The top of the T cast a shadow down the bar, which had a notch for each hour. In the morning, the cross bar was pointed to the east, where the sun rose. In the afternoon, it was pointed to the west, where the sun set. Some shadow clocks were small enough to be carried around.

An Egyptian shadow clock

Early sundials

Sundials were first used thousands of years ago by the Egyptians and the Babylonians. Early sundials were made of hollow half bowls with a raised part in the center to make the shadow. The bowl was divided into 12 parts. Later, sundials usually had a flat surface, divided into 12, like a traditional clock face. They had a fixed, upright pointer in the middle. As the sun moved across the sky, the pointer cast a shadow that moved around the sundial's face, marking the hours.

This huge sundial was built by the Indian ruler and astronomer Jai Singh, as part of his observatory in Delhi. It casts a shadow in a hollow bowl shape.

Did you know?
The word *hour* comes from the Latin word *hora*, which comes from a Greek word meaning season.

Sundials today

Private dials in people's gardens were usually horizontal and were read from above. Public sundials were usually vertical and set high up on buildings for everyone to see. Sundials are still being made today, mainly for decoration, as they are only accurate to within about half an hour. They often have beautifully carved or painted faces. There are even small portable sundials. These may have compasses built into them so that they can be pointed in the right direction. Sundials have to be turned so that the shadow points to the number 12 on the sundial's face at midday.

The pointer on a sundial is called a gnomon. This means "one who knows." On horizontal sundials the gnomon usually points north.

WATER, FLAME, OR SAND

Many people measured time using shadows, but this caused a problem at nighttime, or when it was cloudy. So people began to invent other ways to measure time.

Some of the earliest civilizations developed ways of telling the time using water, flame, or sand. They measured the hours, but did not measure minutes or seconds. Using water, flame, or sand helped people to measure time more exactly than they could by looking at shadows cast by the sun.

Did you know?
Some Romans who owned water clocks also had slaves to call out the time. They were the world's first speaking clocks. In the center of Ancient Rome, there was a slave who called out every day when it was noon.

Water clocks

The Ancient Egyptians, Greeks, and Romans all used water clocks. Water dripped from one container into a lower one which had the water level marked off to show how much time had passed since the clock was started. In Ancient Rome, water clocks were used in court to make sure lawyers didn't speak for too long. But when the Romans took water clocks to some of their conquered lands, they had problems. In the cold weather, the water froze and they could not tell the time.

Candle clocks

Candle clocks were simply candles with lines marked down the side at regular intervals. By looking at how many marks were left, people could tell how long a candle had been burning. Some candles had different smelling waxes in them. As the candle burned down, people could smell what time it was.

Early flame alarm

The Ancient Chinese had the bright idea of making an early alarm clock using a candle, a piece of string with weights tied to both ends, and a gong. They laid the candle on its side so that it burned at a constant rate. Then they put the string across the candle. When the flame reached the string, it burned through, and the weights fell on to the gong.

Sand timers

Sand timers were not used until the fourteenth century. They were also called hourglasses because the sand took an hour to run from one side of the glass to the other. Although they were called sand timers, real sand was too thick, so they used finely-powdered eggshell. In medieval times, preachers often used huge, one-hour timers to time the length of their sermons.

This sixteenth-century sand timer was used on board a ship. It was made in two halves and joined with wax.

Ship time

Ships used to have at least two different sand timers on board. One took four hours to run out and was used to time how long a sailor stayed on duty. The other took 28 seconds to run out and was used to time how fast the ship was moving. A sailor threw a length of rope, with knots tied at regular intervals, over the side. As the first knot passed through the sailor's hand, the sand timer was turned over. The number of knots that passed through his hand in 28 seconds gave the speed of the ship in knots.

This picture shows a candle clock with 14 doors. When the candle had burned for an hour, a ball dropped from the bird's beak. This caused a door to open and a figure to come out. The clock was made in Egypt in about A.D. 1200.

Mechanical clocks were developed in the thirteenth or fourteenth centuries. At first they were mainly used by monasteries and churches.

Time was important in monasteries and churches because prayers and sermons were said at regular intervals each day. Religious life was ordered by the ringing and chiming of bells. Gradually clocks were built in public places such as town squares and marketplaces. These chimed the hours and became an important part of town and village life.

Did you know?

The word *clock* comes from the French word *cloche*, meaning bell. This is because clocks once rang to announce the hours of prayers and religious services.

A Chinese first

The first mechanical clock was built in 1088 by a Chinese official called Su Sung. His clock was made up of shafts and cogs, which were worked by a large waterwheel. Water filled scoops on the wheel which then moved around one spoke at a time. The clock even struck the hours and the quarter hours. The Chinese did not develop Su Sung's invention, and his idea was lost over time.

One-handed clocks

The first European mechanical clocks were not very accurate. They were powered by a heavy weight fastened to a cord, which was wrapped around a drum. As the weight fell, it turned a wheel, which moved gears, which turned the clock hand. The falling weight was slowed by a mechanism called an escapement. One of the oldest existing mechanical clocks was made in 1386, and can still be seen at Salisbury Cathedral in England.

A one-handed mechanical clock made in London in 1653. Minute hands were not invented until 1670.

Counting the seconds

In 1657, another scientist, a Dutchman called Christiaan Huygens, took Galileo's idea further and used pendulums to make the first clocks that were accurate enough to count seconds.

Accurate clocks

The clocks that developed after 1657 used a pendulum, cogs, springs, and a falling weight. Clockmaking became a skillful job and clocks became very accurate. They were built on most public buildings. Rich people collected grandfather clocks and beautifully decorated portable clocks. Some modern clocks are still mechanical. They use springs and cogs and need to be regularly wound up to make them go.

Pendulums

By the end of the fifteenth century, clocks were made with springs rather than weights. But they were still not very accurate. Then, in 1581, Galileo noticed a lamp in the Cathedral of Pisa swinging in a regular pattern. This gave him the idea of using a pendulum, or swinging weight, to time things. He didn't apply his idea to clocks until much later, and died before he finished his work. Leonardo da Vinci had also envisioned a pendulum clock in his sketchbooks in the late fifteenth century.

Time Fact
Horology is the science of measuring time. Someone who makes watches and clocks is a horologist.

WATCHES AND CLOCKS

The first watches were made in the early 1500s. Since then, they have developed from beautiful pocket watches to the inexpensive, accurate, and reliable quartz wristwatches most people wear today.

In about 1500, a German locksmith named Peter Henlein started making clocks using a spring rather than a weight. As the spring slowly unwound, it turned the gears. Clocks had to be wound regularly to wind up the spring again. The use of a spring rather than a weight meant clocks could be made much smaller and could be carried around, which led to the invention of watches.

This beautifully decorated watch was made in 1640, in Switzerland.

First watches

The first watches were made in Germany in 1502. They were about the size of a softball and were carried around in the hand. Often egg-shaped, made from precious metals, and studded with jewels, they were not very accurate. As the spring slowly unwound, it affected the speed of the turning cog. In about 1530 a clever watchmaker from Czechoslovakia, named Joseph Zech, found an answer to the problem. He invented a small mechanism shaped like a cone. This fitted inside watches and made them more accurate.

Wristwatches

Until the start of the World War I in 1914, only women wore wristwatches, while men usually wore pocket watches. But pocket watches were not practical for soldiers in the trenches, so wristwatches were worn by men, too.

Time Fact
Quartz clocks and watches are very accurate, often to one second in 30 years.

Quartz time

Quartz crystal is a special type of mineral that produces minute amounts of electricity if there is an input of energy from outside. In a watch, a small battery provides this energy. The battery causes the quartz to vibrate regularly so that, unlike a watch with cogs and springs, the quartz watch does not need winding up. In modern homes, there is a quartz timing device on most pieces of electrical equipment.

Analogue and digital

Modern clocks and watches are either analogue or digital. An analogue face has hands that move around to show the time. The face is divided into 12 hours. Sometimes the numbers are shown as Roman numerals. From one to 12 these are: I II III IV V VI VII VIII IX X XI XII. A digital display has no hands, but shows the time in numbers on a screen.

The inside of a watch is made from metals that won't rust. Some parts may be made of jewels because they don't wear away.

Mini computers

Some wristwatches are so clever that they are like mini computers. You can find out the time anywhere in the world, or use them as stopwatches to time things. They may have alarms and can be used as calculators.

TIME AND NAVIGATION

Time has had an important role to play in helping sailors navigate safely around the world's oceans.

By A.D. 150, a Greek astronomer and mapmaker named Ptolemy plotted imaginary lines round the world, called latitude and longitude, on the first world atlas. Latitude lines tell you how far north or south you are.

The astrolabe

Early navigators knew how to find their position on the lines of latitude by using a mariner's astrolabe. This was made up of a circular metal ring marked with degrees, and a rotating pointer. Sailors could line up the sun or a star on the astrolabe and measure the height of the sun at noon, or of the star at night. They worked out their latitude by comparing their measurements with those in an almanac, a kind of book that gave the angles of the sun and stars at different times of the year. The mariner's astrolabe was very popular in the fifteenth and sixteenth centuries, but it was only useful when the sea was calm.

This map, made in 1540, was based on Ptolemy's map and shows the lines of longitude (north to south) and latitude (east to west).

The sextant

In the eighteenth century, a device called a sextant replaced the astrolabe and helped sailors measure lines of latitude more accurately. They used the sextant to measure the angle between the sun and the horizon at midday. They knew when it was midday by timing the hours with an hourglass.

Longitude

Longitude lines are imaginary lines drawn around the Earth, passing through the north and south poles, but giving an east or west position. In the past, working out longitude was much harder than finding latitude. Navigators knew that the Earth turns 360° every 24 hours. This means that it turns 15 degrees each hour and 1 degree each 4 minutes. They knew that if they sailed at a certain speed on a fixed course, and compared time back home with time at sea, they would be able to fix their position against the lines of longitude.

Needing a clock

The problem was that to measure their position on longitude, sailors needed a reliable, accurate clock. Sand timers were not very accurate, and the rocking of a ship conflicted with the swinging of a clock pendulum. The heat and cold made spring clocks inaccurate. Not knowing how far east or west a ship was made sailing very dangerous and many ships sank because of this. In 1707 a British fleet of ships led by Admiral Sir Cloudesley Shovell sank near the Scilly Isles—Admiral Shovell thought his ship was just off the coast of France.

John Harrison (1693–1776) spent 40 years making the perfect clock for navigators to use on ships.

A reward

In 1714 the English government offered £20,000 (a huge reward) to anyone who could solve the problem of measuring longitude. John Harrison, an English carpenter and self-taught watchmaker, found the answer where dozens had failed. In 1728 he started work on a special clock called a chronometer that would not be affected by a ship's movement or changes in temperature. In 1773, after his fourth chronometer was built, he was paid the reward money.

Did you know?
When Harrison's fourth chronometer was tested on a five-month voyage to Jamaica, it only lost about 15 seconds.

STANDARD TIME

Did you know?
If you move west from Greenwich, you lose one hour for each time zone. If you move east, you gain one hour. Today GMT is called Universal Time Coordinated, or UTC.

Before the nineteenth century, most towns had their own local time. One of the changes that made places set their clocks to the same time was the arrival of the railroads.

When the railroads were built, especially across the U.S. and Canada, the different times in the different areas of these huge countries caused great confusion. How could train timetables work if everyone had a different idea of what time it was? In 1884, an international conference was held in Washington, D.C. and the nations agreed upon a system of standard time which is still in use today.

Standard time

Standard time uses 24 lines of longitude, each representing one hour's difference in time. The first line of longitude runs through Greenwich in London and is called the Prime Meridian. The time at the Prime Meridian is called the Greenwich Mean Time (GMT) and all other times are measured from this. The area between the lines of longitude are called time zones. Within each zone the time is the same.

The 24-hour clock

Each day begins at midnight and lasts for 24 hours. Clocks using the 24-hour time system show four digits. The first two show the hours past midnight and the other two the minutes past the hour. Midnight is the end of one day and the start of the next, so 2400 on May 25th is the same time as 0000 on May 26th. The 24-hour clock avoids confusing morning and afternoon times.

A scientist is looking at the beam of an infrared laser, part of an atomic clock.

International Date Line

The International Date Line is an imaginary line, not quite straight, which goes from the North Pole to the South Pole. It is on the opposite side of the world from the Greenwich longitude and passes through the Pacific Ocean. This line separates one day from the next. If you cross this line westward, a calendar day is added. If you cross it eastward, you lose a calendar day.

Atomic clocks

Since the 1940s, scientists have developed atomic clocks. They are controlled by the vibration of atoms, and are the most accurate timekeepers in the world. They tick more than 9 billion times a second and are accurate to a nanosecond. An atomic clock is used to give the world standard for the length of a second. Scientists rely on their accuracy when developing programs such as satellite navigation systems.

Greenwich longitude was first measured through the old Royal Greenwich Observatory in London in 1884.

NATURAL CLOCKS

Nature influences the animal and plant worlds, and many things happen for a natural period of time, unaffected by the ticking of our clocks.

Nature's natural cycles and rhythms create what is sometimes called a biological clock. Biological clocks influence when flowers come into bloom and when birds and animals migrate. They also influence when we need to sleep and to wake up.

Time fact
The biological clock of people, animals, and plants has a day between 23 and 27 hours long. This time cycle is called circadian, from two Latin words meaning "about a day."

Flower clocks

In the nineteenth century, gardeners planted flower clocks in the gardens of big houses. They knew that certain flowers opened and closed at particular times of the day. These were arranged like a clock face, in flower beds, to show daytime hours. They could be accurate to within half an hour.

A plant mystery

Many plants open and close regularly according to the time of day. In about 327 B.C., a general in the army of the ruler Alexander the Great wrote that the tamarind tree of India folds its leaves at night and opens them each morning. The first scientist to experiment with the regular opening and closing of plants was a Frenchman named de Mairan who wrote about his work in 1729. Today, scientists are still trying to discover what makes plants' biological clocks work.

The African marigold (far left) traditionally opens at 7 A.M. The passion flower (left) opens at noon. Both were grown in Victorian flower gardens.

Birds use the positions of the sun, moon, and stars to navigate, with great accuracy and over vast distances, to their new seasonal homes.

Migration

The body clocks of animals and birds tell them when to leave one part of the world for another. Scientists think the animals' and birds' internal clocks are triggered by slight changes in the position of the sun, moon, and stars.

Human clocks

How accurately can you guess the time taken by a minute? You need a stopwatch to time this properly. But we do have a biological clock inside our bodies. This clock tells us when to wake up and when to go to sleep. It also keeps our bodies at the right temperature. The human body does not like its biological clock to be disrupted. When people fly a long distance over several time zones, the body becomes confused and the traveler suffers from jet lag.

Did you know?
The expression *tempus fugit* (time flies) comes from the Romans. In fact, they were referring to the weather and the clouds flying overhead.

A.M. These letters stand for "in the morning", from the Latin *ante meridian* ("before noon").

analogue face A clock or watch face divided into 12 parts. The time is shown by the clock hands.

astrolabe An instrument used by sailors to measure the position of sun, moon, and stars.

astronomer Someone who studies the science of planets and stars.

atomic clock A kind of quartz clock where the quartz is powered and controlled by vibrating atoms.

book of hours A book that gave the time for prayers and services. The books were often beautifully illustrated with pictures of the seasons and months.

calendar A system for measuring time over a long period, usually a year or a month. Some ancient calendars were based on the Earth going around the sun, which takes a year. Others were based on the moon going round the Earth, which takes a month. Modern calendars are based on a year, broken down into months and days.

chronometer An accurate device for measuring time, usually on board ships.

circadian The internal natural clock of plants and animals. It comes from a Latin word which means "about a day".

decade A period of 10 years.

digital face A clock or watch face that has no hands, but displays numbers to show the time.

fortnight A period of 14 days.

Gregorian calendar Our modern calendar, named after Pope Gregory XIII. It was started in 1582.

horologist A watch or clockmaker.

horology The science of measuring time.

hour A period of 60 minutes. There are 24 hours in a day.

hourglass A timer made of two glass containers joined in the middle. Sand takes one hour to run from one end to the other. Modern sand timers may take only minutes.

International Date Line The line of longitude which divides one day from the next.

Julian calendar The first modern calendar, named after Julius Caesar and created in about 45. B.C.

latitude Imaginary lines that circle the Earth parallel to the Equator.

leap year A year that has 366 days in it. There is a leap year every fourth year.

longitude Imaginary lines that circle the Earth passing through the North and South Poles.

lunar month The average time the moon takes to circle the Earth. Ancient calendars usually used 29 or 30 days for a moon month.

midday 12 o'clock in the middle of the day. On the 24-hour clock, midday is 1200 hours. It is the time when the sun is at its highest point.

midnight 12 o'clock in the middle of the night, when one day changes to the next. On the 24-hour clock, midnight is 0000.

millennium A period of 1000 years. It also means a thousandth anniversary.

minute Part of an hour. There are 60 minutes in an hour. Minute hands were invented in 1670.

nanosecond A very exact measurement of time used by scientists. A nanosecond is one thousand-millionth of a second.

noon Another name for midday. The word comes from the Latin for ninth hour. In Roman times, it was the ninth hour from sunrise.

observatory A building used by astronomers to measure the movements of the stars and planets.

pendulum A swinging weight on the end of a bar or string used to drive clocks.

P.M. Letters standing for "in the afternoon" from the Latin *post meridian* (after noon).

season A year can be divided into seasons. Some countries have four seasons called spring, summer, autumn, and winter. Other countries may only have two, called the wet and dry seasons.

second Part of a minute. There are 60 seconds in a minute. Clocks could not count seconds until 1656.

solar year A solar year is the time the Earth takes to circle the sun. This is usually said to be 365 ¼ days. Actually it is 365 days, 5 hours, 48 minutes, 45.5 seconds.

stopwatch A type of watch used to time short amounts of time accurately. The second hand can be stopped very quickly.

sundial An instrument used to tell the time by means of the sun's shadow.

time zone An area between lines of longitude in which the time is the same.

year A period of time of 365 or 366 days in the Gregorian calendar, starting on January 1.

watch A clock small enough to be carried, usually in a pocket or worn on the wrist.

INDEX

ML   4/02